Book 1

An Introduction to

The Science of Offsetting

Jamen Zacharias

ATI Publishing

Lytton B.C. Canada

An Introduction to The Science of Offsetting – Book 1

By Jamen Zacharias

Copyright @ 2013 by Jamen Zacharias

Published and distributed by ATI Publishing.

www.jamenzacharias.com

All rights reserved. No part of this book may be reproduced, stored in a retrieval system or transmitted in any form or by any means, electronic, mechanical, photocopying, recording or otherwise, without the prior written permission of the author. Brief quotations of one hundred words or less may be reproduced in articles, reviews and books provided the title, author, and copyright date are acknowledged.

Author photograph copyright c Jamen Zacharias

All illustrations copyright c Georgia Lesley

Disclaimer: The author and publisher of this book are NOT RESPONSIBLE in any manner whatsoever for any injury which may occur through reading and following any of the instructions or activities described in this book. It is strongly recommended that you consult a physician regarding your physical conditioning prior to attempting any technique described in this book as the activities, physical or otherwise may be too strenuous or dangerous for some people.

Printed in the USA

ISBN 978-0-9919044-0-2 (pbk. : book 1)

1. Martial arts. 2. Center of mass. 3. Equilibrium. 4. Physics. 5. Sports sciences. I. Title. II. Title: Science of offsetting.

GV1101.Z33 2013   796.8   C2013-902516-2

First Printing - 2013

An Introduction to the Science of Offsetting
Book 1
Table of Contents

Dedication i
Acknowledgements ii
Foreword: By Farshid Kazemi iii
Preface: Why This Book and Why Offsetting? v
Introduction vii
What this Book is NOT! x
Chapter 1: Science of Offsetting 1
    Inner Fundamentals 1
    Pressure Energy and Intention Reading 2
    Controlling Center / Center Line 2
    Why Offset? 3
    Offsetting 3
    Type 1 Offsetting Characteristics: Oncoming Traffic 3
    Type 2 Offsetting Characteristics: Earthquake 4
    Type 3 Offsetting Characteristics: Missing a Step 4
    Type 1, 2, 3 Characteristics: Harmonious Interplay 5
    Riding the Bull 5
    No Quick Fix 6
    Basic Use and Manipulation of Force Distinctions 6
    An Overview 6
Chapter 2: Moving Along 9
    Categories of Pushing and Pulling 10
    The Connection to Reading Pressure 11
    Flow Hands 1 12
    No Grip Shifting 12
    Distinctions in Pushing and Pulling: A List 13
    Inner Realms 14
    More Thoughts about Offsetting 15

- Key Factors in Offsetting 16
- Commentary on Subtle Dimensions of Offsetting 18

Chapter 3: And Furthermore 21
- What Makes Close Quarter Martial Technique Work? 21
- Common Pitfalls? 22
- Offsetting: Applied to Outside Ranges 25
- How Do We 'Provoke' Movement without 'Invoking' Escalation? ... 26
- Three Keys to Moderating Force 27
- The Need for Hands On Training and Further Study 28

Chapter 4: Close Quarter Arts & Distinctions 31
- Conventional Clinch 31
- Modified Clinch 31
- Circle of Space 32
- Flow Hands 2 32
- Shifting Arts 33
- Locking Arts 33
- Trapping Arts 34
- Striking Arts 35
- A Note 35

Chapter 5: Applied Jeet Kune Do Philosophy 37
- Stages of Cultivation 37
- Partiality 38
- Fluidity 38
- Emptiness 39
- The Fitting in Spirit 40
- Stages of Cultivation Applied to Offsetting 40

Closing Comments 43
The Symbol of Circles and Lines 44
About the Author 45

Dedication

"I offer this book in the spirit of furthering the advancement of cultures, arts and sciences of justice, well-being and peace.

The heart of this book is dedicated to humanity and our shared Divine Luminaries."

"This book is also lovingly dedicated to my beloved wife, Sooyeon Zacharias, a loving and devoted partner. She is an inspiring yoga and health & wellness educator working diligently to harmonize cutting edge sciences with ancient wisdom. She is a constant source of inspiration to me and so many others."

Acknowledgements

"I wish to extend my sincerest thanks to my wife, all of my family, friends, students, All Things Institute instructors and all of my teachers and mentors. I have been truly humbled by the tidal wave of love and support I have received from so many, without which I feel none of my pursuits would have come to fruit."

Illustrations: Georgia Lesley (Professional Artist & Illustrator)

Foreword: Farshid Kazemi (Scholar and Poet)

Editing: Vicky Hansen, Georgia Lesley, Mary Jane Oakes

Book Formatting: Mary Jane Oakes (Oakes Web & Media)

Book Reviews: Brenda King
Shahram Moosavi
Shishir Inocalla

My Teachers: Sifu Thomas Cruse, Datu Shishir Inocalla, Sabu Mr Min, Sifu Paul Vunak, Sifu Michael Gruener, Master Tortal and Sabu Mr Jong

Important Influences: Bruce Lee, Dan Inosanto, Donn F. Draeger, Morihei Ueshiba, Gioia Irwin

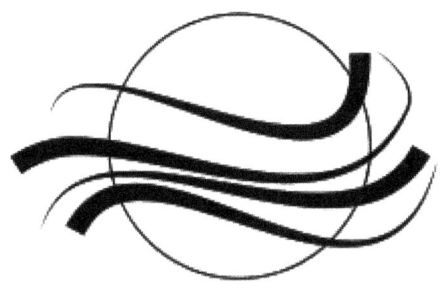

Foreword

The lion who breaks the enemy's ranks is a minor hero compared to the lion who overcomes himself.

~ Rumi[1]

The task of introducing this marvelous booklet is not an easy one. Since I am not a martial artist, I am neither capable nor qualified to comment on the technical aspects of this art and science. However, since I have a personal and deep friendship with the author, and my scholarly training resides in matters pertaining to religious studies and literary theory, I will speak of the Spirit (or put in another way, the Tao), that animates this new art and science, and to which I have had the singular and profound privilege to witness.

Every student of the martial arts who has made a deep study of the particular art to which they have attached themselves, will soon discover that indeed, the spirit behind the art in which they are practitioners, draws its inspiration from a spiritual tradition or religion, whether it be Hinduism, Shintoism, Buddhism, Taoism, Islam, to name but a few. The art and science ably discussed in these

---

[1] Camille Adams Helminski, *Rumi Daylight: A Daybook of Spiritual Guidance* (Boston Massachusetts: Shambhala, 1999) 40.

pages, also owes its spirit to a specific spiritual tradition, which though new in the annals of history, embodies within itself that primordial Spirit, which is at the origin of all spiritual traditions. Indeed, there is no doubt that were it not for this Spirit, this art and science would have ceased to exist, for its practitioner needed to see beyond the outwardly seeming violent façade of the martial arts, to penetrate into its inner peaceful reality. Indeed the overcoming of this dichotomy, this binary opposition between the violent past of the arts and its peaceful future potential, is at the heart of this new art and science. There is no need here to state what this religion or spiritual tradition is, or what its name, teachings, or principles are, for in the age of "Google" anyone interested in the biography of the author of this booklet, may well soon discover it. In the end, just as Bruce Lee – who is one among the various genealogies of this art and science – states, "It is like a finger pointing away to the moon: Don't concentrate on the finger, or you will miss all that heavenly glory." Thus, if one concentrates upon the finger, which is the outward, physical, and material aspect of the art, however necessary, one will miss all the "heavenly glory" of the moon, which is the inner, mystical, and spiritual reality of this art and science. But, what this Moon is, or more precisely Who this Moon is – is the answer to be sought by every true practitioner, nay every human being.

It is my hope that this little finger (book), may in its own way, point to the splendour of the Moon.

Farshid Kazemi

University of British Columbia.

January 30, 2013

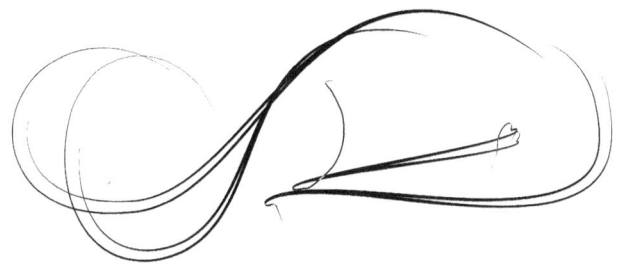

Why This Book and Why Offsetting?

Over the years, I have delved into many facets of martial arts study. All were of merit, yet none of them stood out for me enough to personally invest the energy, time and resources needed to produce a book, let alone a series of books. This was not until I deeply considered the science of offsetting.

Anytime that I have published an article on a martial arts subject, the science of offsetting is at least mentioned as being a focal point from which those subjects are best actualized. Therefore, this book is a humble attempt at producing something of value within the collective research of martial sciences and to hopefully contribute in some manner to the advancement of its future culture.

When considering the multi-facets that are possible in martial arts, what lies at its root can often become clouded. We can become caught up in the plethora of varying traditions, techniques or styles, rather than digging deeper into what is truly at the root of it all.

Mastery of these skills requires we understand its depth, from its simplest level. It is an interesting paradox, that to function from the depth, requires we have a profound understanding of its basis in simplicity.

The science of offsetting has been utilized within many martial art styles and systems, throughout the ages. It is my intention however, to expose more of it from its nameless, formless root. The reason for this is to be able to look upon it with fresh eyes and to invite a progressively deeper understanding. As we enter a new age, humanity has gained a chance to revisit old ideas, and in turn reshape them to fulfill the needs of a stimulated spirit of fresh insight.

The arts of war have been perfectly honed, yet the arts of peace have had virtually no serious consideration. I present this book with the idea that the elements that make up these sciences become considered from enlightened perspectives and noble ideals, while not falling short of the true, profound beauty of honest functionality.

Introduction

The ability to masterfully 'offset' (the calculated disruption of the recipient's center of gravity and sense of balance) provides you with a nucleus that is surrounded by every degree of force capable of being generated from within you.

Keep in mind, offsetting is on its own, non-lethal.

This means that one's physical manipulation of force and energy is now governable from the ground floor, upwards.

This provides the ethically minded the opportunity to acquire skills that are tightly bound to justice and exercisable according to mercy, rather than chained to the predictable, usually less than noble results of brute force.

It is a science that is capable of mirroring one's temperament, therefore challenging those that wish to emanate peace, to develop further both externally and internally, so that loss of temper, misplaced competitiveness or retaliatory impulses, may be kept safely in check.

While not limited to any style offsetting is prominent within virtually every martial art worldwide.

Not having this element of skill identified and highly developed, robs the practitioner of the plethora of potential that is provided in innumerable techniques, paradigms and art forms.

It does not matter which art of physicality you practice, without the epicenter developed, we are bound to merely mimic what others have perfected.

Martial Arts Distinctions:

I wish to layout a series of lists containing distinctions of martial arts practice. Without going into detail, you may notice that everything upon the lists has their own unique focus, demanding differing modes of practice that would require entirely distinct approaches to understanding and teaching.

Each can be subjected to varied perceptions and will be understood according to a myriad of maturity levels.

A General List:
    a. Special Operations Forces
    b. Army and Navy Forces
    c. Police & Peacekeeping Forces
    d. Personal Protection
    e. Sports and Competition
    f. Esthetic Form and Beautification
    g. Health and Longevity
    h. Esoteric Arts
    i. Cultural and Traditional Preservation
    j. Entertainment, Stunt and Demonstration
    k. Child and Youth Education and Development
    l. Mental Health and Therapy
    m. Formlessness, Adapting Applicable Elements from all of the Above

Initial Training Stages of ANY of the ABOVE:
    a. Learning Technique/Strategies/Principles
    b. Practicing Technique/Strategies/Principles
    c. Mastering Technique/Strategies/Principles
    d. Functionalizing Technique/Strategies/Principles
    e. Maintaining Technique/Strategies/Principles

Training Stages May Fall Within:
    a. Isolated Practice and Personal Contemplation
    b. Varying Forms and Degrees of Partner Drilling
    c. Varying Forms and Degrees of Sparring
    d. Varying Forms and Degrees of Competing
    e. Varying Forms and Degrees of Conditions and Scenarios

Distinctions in Forms and Degrees of Conditions & Scenarios:
    a. Self Defense

b. Life Preservation
   c. Intervention
   d. Transporting
   e. Controlling

Further Distinctions:
   a. Outside Ranges
   b. Inside Ranges
   c. Stand Up
   d. Ground
   e. Multiple Attackers

Even Further Distinctions:
   a. Edged Weapons
   b. Impact Weapons
   c. Improvised Weapons
   d. Environmental Weapons
   e. Traditional Weapons
   f. Projectile Weapons
   g. Wipes, Chains
   h. Firearms
   i. Explosives

And More Distinctions:
   a. Ways of Intercepting Movement and Intention
   b. Ways of Deceptive, Non-Telegraphic Movement
   c. Ways of War and Stealth
   d. Ways of Non-Violent Communications
   e. Ways of Non-Violent Posturing and Movement
   f. Ways of Peace, Scientific Insight and Application of Divine Precepts

Offsetting is not confined to any martial style or particular mould.

This book introduces an analysis of what makes up the science of offsetting. It may in turn be applied according to the needs dictated by the relevant circumstances or situation.

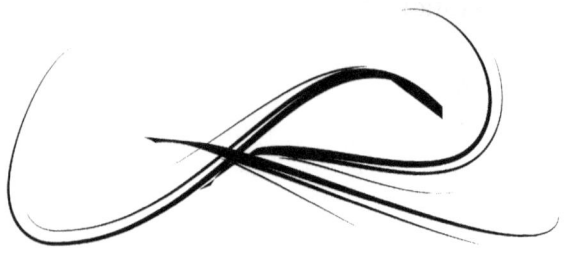

What This Book is NOT!

When discussing martial arts in this day and age, it is easy for those interested to assume the differing martial art disciplines share one and the same goal, that being self defense.

This is why I wish to be clear that this book is not a self defense training manual. It acts merely as a component to the basic study of the sciences being explored herein. Of course the principles and knowledge shared within these pages is very relevant to the overall mastery of the self defense arts, however this book is not offered as a quick fix towards obtaining self defense skills.

This book is really only a small contribution to a much higher learning and far greater dialogue. It is aimed towards those that teach the arts and those interested in breaking down elements involved in the ethical mastery of the manipulation and control of force and energy.

Chapter 1: The Science of Offsetting

Within these pages I am striving to describe what I feel are some of the inner mechanisms that make our skills and use of technique expressible according to a certain standard.

To me, the personal mastery of the manipulation of force demands refinement of certain fundamentals, alongside and to be expressed within, one's physical training.

Inner Fundamentals, the minimum of which are:

a) Avoiding triggers of impulse and escalation in our self and the recipient.

b) A developed inner condition of calm, composure and proper perception.

c) A receptive spirit and awareness of one's constancy or slipping from the ideals of justice, mercy and nobility.

d) A direction towards functionality, achieved through the minimum amount of effort.

With these principles at heart, I offer a breakdown, though not comprehensive, as it is beyond the scope of this book, of some essential components regarding personal manipulation of force, within close quarters.

∼

Pressure, Energy and Intention Reading

The reading of pressure (while in cohesion/contact with another person's limbs/body) is indeed of vital importance.

To 'read' is the ability (while in physical contact) to properly interpret and monitor the pressure, energy, lines and intention of the one whom you are in close contact with, as well to be able to feed, utilize and dissolve this pressure and energy.

It is the foremost key to being able to connect with the recipient's centre of gravity. Through training to develop the ability to read pressure, one acquires an internally felt and externally sensed force intuition. This force intuition in turn increases the accuracy of one's ability to govern the appropriateness and adequacy of one's use and manipulation of force.

∼

Controlling Center/Centre Line

The reason we learn to read pressure and energy is essentially so that we can connect to a person's centre of gravity. Once we have connected to a person's centre of gravity, we have the elements required to 'offset', to take their balance. To 'offset' means to disrupt, upset and control a person's centre of gravity. In addition, this has a marriage of sorts to occupying the person's center line.

Center line is the line of gravity that exists from the top of the body to the ground and it is what is maintained in order for a person to stay on their feet. If a person is most concerned with their own footing, they will be less able to effectively express their own intentions, such as to strike, control you or physically lash out.

∼

Why Offset?

Again, the essence of 'offsetting' is to take control of a person's centre of gravity, therefore disrupting their sense of balance. When we have disrupted a person's sense of balance, the success of their own intentions becoming physically expressed is greatly reduced.

Certain forms of offsetting can also be done in a manner that creates a sort of 'state of bewilderment' or 'temporary confusion' in the recipient. This induced state is more conducive to potential de-escalation, as it moves the recipient from a common reactive, impulsive mode or behavior, to a confused, yet nonetheless 'thinking mode', which by its nature and reality is less impulsive.

Offsetting

The term offsetting means for our purposes, to disrupt the recipient's center of gravity, to take their balance. Offsetting can be employed in a manner that results in both physical and mental disruption.

We have broken the essence of offsetting into three types or characteristics. It should be mentioned at this point, that the manner in which we are offsetting, is not based in throws, hip tosses, sweeps or the like. What we will be describing (the reader can imagine), involves the recipient still being in an upright position.

This does not mean we are for or against downing a person in some fashion (which is an important part of offsetting), but merely to help the reader realize that what is occurring to the recipient is happening while they are standing up. The takedowns are a choice, a follow up, based on a case by case situation.

Type 1 Offsetting Characteristics: Oncoming Traffic

A Type 1 offset mirrors the characteristics of being struck by a car, or what we call being struck in oncoming traffic. It is not a pleasant image. However it does make the point quite well, as one can imagine the force of being struck as being capable of moving the recipient in whatever direction the 'oncoming/going force is headed. This type of offset can be abrasive, yet is not confined to being so, and is basically a committed, linear energy. The force of movement follows through the recipient in such a fashion, that the 'whole' body moves in the direction of the force driving it.

∼

Type 2 Offsetting Characteristics: Earthquake

A Type 2 offset mirrors the characteristics of being caught up in an 'earthquake'. An earthquake sort of bumps, tips and shakes, from one direction to another in often quick, jolting manners causing the recipient to lose their sense of balance and equilibrium.

This type of offset can be performed with a high level of explosive, percussive energy, but again is not confined as such, for at the higher levels of skill one may employ more subtle offsets by 'complimenting' energy, force and direction. With this level of Type 2 offsetting one is able to 'capture momentum' and 'compliment it', guiding the recipient, in a non-forceful manner, to 'adjust' according to the flow of their own energy.

These types of offsets usually require a re-wiring of sorts to our thinking, as we must be able to bump and tip, without over muscling it. When we can do this, the chances of a reactive, impulsive response to what we are doing is greatly reduced.

∼

Type 3 Offsetting Characteristics: Missing a Step

A Type 3 offset mirrors the characteristics of missing a step on a flight of stairs or expecting some form of resistance, only to have it completely dissolved. This is a more subtle version of unbalancing and controlling the centre of gravity than the first two types of offsetting. It is caused by an absence of force, rather than by force being directly applied or through complimenting it. This type of offset is usually followed immediately by a Type 1 or 2 offset.

Type 1, 2, 3 Characteristics: Harmonious Interplay

Each type of offset is meant to work in harmony and concord, in conjunction with the other, creating a state of perpetual falling if you will. In the continuum of practice, it is almost dance like. When training these three types of offsets it is important to desire a level of effortlessness, by allowing for the characteristics of each to have their due course. This is achieved by not focusing upon any one of them for too long.

The full potency of all three types of offsets working together is that the recipient experiences a feeling of complete loss of center, while still being controlled in a manner that is potentially trustworthy (by remaining upright). These types of offsets require a moderated use of energy, rather then a bullied, brute force expression of strength. If this is achieved, the triggering of the impulses of escalation or retaliation will be potentially reduced.

A suggestion in this regard is to learn to moderate one's energy and force, to 'choke it off' if you will, by allowing momentum to occur, rather than just continually exerting your own force.

Principles such as these are not easily explained without the opportunity to feel their results. This type of skill is not based in the esthetic appeal of an audience, but in hands on experience. That is where the appreciation is truly gained for its potency and potential.

∽

Riding the Bull

A common mistake is when we feel we must be doing something all the time, which leads to muscling our way through technique and wastes of energy. I feel it is the hallmark of a mature practitioner when they are able to rest, relax and not need to do anything for a time, while riding out the energies being imposed upon them.

When we are with someone that is able to control us physically, by brute force or higher skill, it is wise to know when to 'ride the bull' of their actions, as opposed to trying desperately to gain control by implementing our own. Attempting to implement our own force at the wrong time is a potential 'trigger of escalation'.

We must learn to discover the appropriate time to insert our own force, or to compliment the force being imposed upon us, while guiding the exchange of energy to an appropriate end of our own design. These suggestions sound idealistic, yet it is my experience that it pays off to truly follow these ideals through, so that the appropriate skill sets begin to emerge from within us, while remaining consistent with these ideals.

No Quick Fix

The idea of doing what it takes to cultivate a true level of mastery may be getting lost in the attention span dwindling construct of modern society. Though I believe we should streamline the learning process of our crafts and sciences, the same realities for mastery are required: time and intelligent effort.

Basic Use and Manipulation of Force Distinctions

1) Force to Cause Movement

2) Force to Control Movement

3) Force to Cause Pain

4) Force to Cause Injury

The science of offsetting, though it can be employed to enact any one of the above force distinctions, and in various combinations with each other, on its own is based in causing and controlling movement.

An Overview

The science of offsetting is closely tied to the high level ability of reading pressure while in contact with another person. Though it is an art that requires real training and proper instruction to acquire, it is a skill set that renders one more capable of appropriating one's use of force and energy.

This is further enhanced by considering the whole entity, which includes breath control, reflection and meditation as well as an active practice of moderating one's use of energy. These additions serve us well by relieving the impulses of brute strength, in its place, the desire to move with ease, in a mindful and effortless inner zone.

*Chapter 1 is an updated and revised article, written by Jamen Zacharias, previously published within Michael Joyce's online magazine www.combativecorner.com *

Chapter 2: Moving Along

Thus far we have discussed insights regarding the fundamentals of what makes up the core of offsetting, describing Type 1, 2, & 3 offsets, termed Oncoming Traffic, Earthquake and Missing a Step, along with Harmonious Interplay and Riding the Bull. We then concluded by considering the relationship or marriage between offsetting and reading pressure, energy and intention.

As previously stated, the ability to read pressure cannot be over stressed, as it is the core skill leading to the ability to offset (controlling/disrupting a person's center of gravity). Controlling a person's center is the foremost goal, as it is what enables us to appropriately implement force, according to the needs of the circumstances, drawing from a more complete scale of potentialities.

When we offset, the sensation to the recipient is not confined to a physical disruption, but invokes a mental disruption as well. It is the combination of these that make offsetting so powerful. We must fall to the simplicity of it and realize the potency of offsetting, as it is a fundamental pursuit for anyone to regain their sense of balance, once it has been lost, before all else.

Essentially, the basic concept of offsetting is in the simplistic skill set of pushing and pulling. Again, the vision that should be in the mind of the reader is that we are 'offsetting' while the recipient remains upright. At this point we are not discussing takedowns or the like.

While sounding purely natural and unsophisticated, pushing and pulling can indeed reach high levels of intelligence and sophistication. It is from the ideal of these higher levels of sophistication that we give the sciences of pushing and pulling the name, the 'shifting arts'.

∼

Categories of Pushing and Pulling

There are basically two core categories or characteristics to pushing and pulling.

To the Whole: When 'pushed or pulled' the 'whole' body is significantly affected.

To the Portion: When 'pushed or pulled' a 'portion' of the body is significantly affected.

Whole: A good example is when you push the centre of the chest. This causes the body as a 'whole' to be moved at once. Therefore pushing to send or pushing for continuous pressure is common when pushing the 'whole' body weight or centre of gravity.

A helpful image in this regard, one that I find very effective in my own expression of offsetting, is that the core structure of the body is its spine. In essence, when we control an upright person's centre of gravity, we are 'moving the spine'.

Portion: A good example is when you push the inner shoulder, on one side only. This causes the body's position to shift, however does not cause the body's entire weight to move at once, as a whole. Only a 'portion' of the body is significantly affected, therefore pushing for position may occur in that area.

Imagine pushing one shoulder, rather then the chest or both. The image that comes to mind is like spinning the body on its axis, rather than moving it from one point to another.

∼

The Connection to Reading Pressure

The concept of moving **the portion** or **the whole** in an intelligent and purposeful manner requires delving a little deeper into the ability to **read pressure**. When dealing with the limbs, the ability to read whether we are 'connected' to the recipient's centre or not, relies on the skill of reading pressure, hence the need to acquire the skill. The reason for this is that pressure in the limbs can change.

The limbs are active and inactive entities and in a way, can become somewhat independent of the whole body when at rest or doing lighter activity, then in an instant, they can stiffen up or resist, becoming connected to the body's 'whole'.

Our ability to 'read' that shift in pressure is imperative when attempting to offset the body through a connection to the recipient's limbs. How we distribute, utilize and feed our own energy depends on our proper interpretation of the energy and pressure within the recipient's limbs.

This acquired ability to read, is a developed skill that directly links the functions of the body to the state of the mind. The mind must be at ease and able to act without hindrances from the body's lower impulses or desires. When training these elements, it is important to find a place internally, a 'zone' that is not caught up in overanalysis.

When we are caught up in the 'analytical mind' our ability to function skillfully is greatly reduced. An indication of a mature practitioner is when one has reached a stage in their studies and practice where the mind is no longer analyzing movements, techniques and strategies and the body is free to act according to the needs and flow of the circumstances themselves.

This stage of skill and 'being' is well beyond the stages of primitive impulses and instincts, as well as the many stages of internal analysis, which consists of learning to implement knowledge in a manner that eventually transcends the impulses and instincts common to the body and ego.

Flow Hands 1

Methods such as sticking hands, pushing hands, pulling hands, percussive hands, passing hands, while learning to put it all together with functional transitions becomes what we call 'flow hands'. The common 'nucleus' of flow hands is primarily 'sticking', with a continuum of transitions, inserts and flow between the other qualities and lines of play.

Cohesion or 'sticking' is the easiest characteristic to achieve and requires no co-operation. The other elements are what I consider more of an insert, based on the needs of the moment or to bring about dynamic practice.

A continuum of all of these qualities, changing and morphing in and out of each other is important, which makes flow hands practice vital, as it is a method that serves this aim.

When we connect, we must lose our 'self' in that connection, whether to the limbs, body or neck, and enjoy the energy we gain from the cohesion that occurs between us. Our centre of gravity can literally become 'one' with the others, and from there, very little effort is required to move through the others' centre of gravity, hence moving them in a manner as effortless as moving our own body weight.

When we achieve this, the other moves as if a part of us and potentially according to our own implied direction. The ebb and flow of this connection is to not be overthrown by competitiveness or a desire to display, but to let the interplay flow to a natural end, guided by our intentions, which should be just and appropriate to the situation. Though this sounds like it takes up an enormous amount of time, the dynamics being discussed can actually occur within a fraction of a moment.

No Grip Shifting

Flow hands practice has been commonly utilized as a flow or continuum to set up other arts such as pushing and pulling with grip, joint locking, trapping hands and striking. If one wishes to reach greater, much more subtle heights in offsetting however, we must forgo the temptation to run to these techniques for a time. In place of the common, we acquire an expertise in bumping, tipping and disrupting the recipient's center of gravity, while in contact with their limbs, without utilizing grip.

I call this skill, No Grip Shifting. It is a skill that requires an enhanced sensitivity to pressure and energy, and a moderating of one's use of force. When we remove the use of grip during practice, when attempting to push or pull, we will enhance our level of cohesion, and learn to moderate and distribute our energy in such ways as to not draw upon brute force. I highly suggest putting the time in here, as we will discover more speedily the need to rely upon momentum and moderation, over brute strength and resistance.

∽

Distinctions in Pushing and Pulling: A List

When employing Type 1, 2, 3 offsets either in an isolated fashion or in a harmonious interplay, there are several characteristics that may be occurring as far as how the recipient may be pushed or pulled.

The following is a list and breakdown of various characteristics:

To Send: To send essentially means to move the body or 'send' it a distance from our own. When one sends it is a clear example of a Type 1 offset, as the 'whole' body is being affected.

Continuous Pressure: Continuous Pressure essentially means to continue to bring a forward pressure (usually) upon a recipient's body, while remaining in contact. This form of pushing is also a good example of a Type 1 offset, as the body as a 'whole' is being affected.

To Create Space: Within close quarters, between two bodies, when one pushes to 'create space' essentially they are expanding upon a condensed characteristic, into a more open one. This form of pushing may be achieved with or without an offset, however when it is achieved in conjunction with offsetting, it is usually a good example of a Type

2 offset, drawing upon more bumping and tipping like energy, while expanding outward.

For Position: When pushing or pulling for position, one may be pushing the portion or the whole, which entirely depends on the needs of the circumstance. To push for position is essentially to adjust the body in some manner in order to achieve another position that is more conducive to achieving a desired end. This is also usually a Type 2 offset.

For Shock: When pushing or pulling for shock, the purpose is truly to disrupt the recipient's equilibrium. It is a type of shock that can result in a disruption that is both mental and physical. This is an explosive, percussive use of a Type 2 offset, yet can result in a Type 1 offset as well.

To Compliment: When pushing or pulling to compliment, the purpose is to assist an already existing force and direction. We are going with the momentum and direction of the mass. This may be a Type 1, 2 or 3 offset.

Resistance Chambering: When resistance chambering, we are either eliciting a degree of resistance or resisting an already existing force and direction, in order to release it. When we release the 'chambered resistance', we create the offset. This is usually a combination of Type 2 and 3 offsets. This can be done against strong energy and resistance and can also been done very subtly, against very small, almost imperceptible degrees of force and resistance. I have found great results from applying very subtle uses of resistance chambering, but it is a fairly high skill, in need of cultivation over time.

∽

Inner Realms

In long term training in these principles, our ability to control someone's movement, while monitoring our own should become second nature. When guided by a higher ethic and not bent on someone's destruction and considerate of the circumstances, we begin to operate in a manner that we are worthy of. Justice and nobility is the light that illuminates from within.

When we are in contact with 'negative forces' of aggression or misplaced competitiveness, we must strive to avoid mirroring those qualities. Our goal instead, is to increase the qualities that are absent (within ourselves or within our opponent), whether in an altercation or in practice.

In basic terms, if someone is fighting with you, you must resist the urge to fight back impulsively, but guide the 'feel' of the interaction to a higher realm of potential, not limited to the excessive, retaliatory norms of our lower tendencies. This cannot be achieved if we are merely matching brute force for brute force, aggression for aggression, retaliation for further retaliation, etc. Acquiring these abilities is not merely a physical thing, but an increase in mental and emotional control. When we are training, indeed when we are living, we can be mindful of this, seeing it manifest creatively in our arts and practice.

Because I wish to address in some depth the subject of one's inner condition within books 2 and 3 of the series, I simply state that one's inner condition quite literally sets the tone for our skills expression. I believe that without acquiring an enhanced inner state, performing offsetting at the highest levels will not be experienced, as one's lower tendencies will take over all too easily, hence drawing upon brute force or aggression, rather than subtlety and enhanced composure.

It is my belief that human beings are purposed to draw upon ever higher qualities, to progressively rely less upon the primitive impulses and instincts of the body and emotions. When striving to utilize spiritual characteristics, our normally hidden potential, eventually becomes exposed and available to us, thus providing completely different characteristics of engagement.

More Thoughts about Offsetting

The ebb and flow of controlling another's center of gravity is what we are after. To be able to offset with relative ease, leaving out the common lower tendencies of brute force and excess is the basic goal. Offsetting in a 'harmonious interplay' is in fact a flow of reading energy and appropriately applying energy, by directing it in ways that match the need.

To me, practice should feel like a standing flow. We want to understand the various distinctions of attributes and qualities that are available so that we do not get tied up and stuck at a standstill based in raw strength and primitively based resistance.

The skill of offsetting is a key to inserting appropriate uses and degrees of force, within the arts of close quarters, such as locking and trapping. In addition, when one has expertise in offsetting, the employment of the striking arts becomes potentially much more potent and effective.

It truly does take a training of the mind, allowing for feelings of how to enjoy the process to the level of full on actualization. It is my suggestion to isolate the three types of offsets, and allow them to come alive through many hours of practice.

When with a qualified teacher, the exploration and flow between the conventional clinch, modified clinch, circle of space and flow hands is the nucleus for how we obtain a good footing in offsetting as we practice it in close quarters. This book should, in my opinion, be studied regularly, read and reread so that the basic elements become fully grasped alongside our regular training and practice.

∼

Key Factors in Offsetting

The following is a breakdown, of key elements for utilizing offsetting, in a manner that is not based in brute force or raw strength. These elements may be employed within virtually any technique capable of disrupting the recipient's center of gravity.

Movement to Momentum: To move (ourselves and the recipient), with the momentum of the ongoing or oncoming force is a key to effortless power. When acting in conjunction with momentum, we are no longer in need of brute force and able to utilize the natural pull of gravity. This must be done in harmony with connecting to the recipient's center of gravity.

Moderation of Force: One of the biggest pitfalls is the excessive use of force. To avoid this, we must become mindful of our force degree, avoiding the impulsive competitive desire to utilize raw strength. There is a time and place for raw strength, yet it should be governed and tapered off during practice and personal cultivation. Raw strength should be cultivated to the extent that even an emergency benefits from a higher expression, rather than a fall back into lower impulses and behaviours.

Choking it off: The moderation of our force is best achieved by doing what I call, 'choking it off'. When we reduce our use of energy we will gain a greater sensitivity to the subtlety of movement. This occurs by 'choking off' our pushes and pulls, part way through an otherwise muscled move. This is also the way to gain momentum and begin to relate to that momentum and movement.

Complimenting Force: To complement is essentially borrowing an already existing or caused force and direction. We are helping it along so to speak, but in a manner nonabrasive or excessive.

Going with the Force: It is a tendency, a lower one, that when we are pushed or pulled, we react through resistance. Go with the force, learn the art of non-resistance.

Shadowing: Shadowing is a paramount skill when used with offsetting, basically, to 'shadow' means to move or compliment the movement and direction of the recipient's body, while staying in close proximity or light contact with that body.

At the same time, while shadowing, we position ourselves, to 'avoid the square off' if possible while occupying centerline and their center of gravity.

Shadowing is a form of relating to the recipient's movements. It is not a mere following, but a relationship to their movements in a manner that remains as safe and advantageous to us as possible.

Shadowing is the 'glue' that makes up a mature ability to interact in a 'Harmonious Interplay', and may draw from elements of Type 1, 2 and 3 characteristics of offsetting.

Shadowing is also a flow of footwork that relates to the movements and direction of the recipient. When we are shadowing, we are also positioning ourselves by slightly zoning to the recipient's side (either one of them, front or back) and maintaining a connection to their center of gravity and centre line.

If I were to organize in simplicity the flow of offsetting, it would include a harmonious interplay of the following forces and skills:

1) Reading, Feeding, Utilizing and Dissolving Pressure, Energy, Intention and Lines of Interaction.

2) Connecting to the recipient's center of gravity.

3) Offsetting (mindfully disrupting their sense of balance).

4) Shadowing (relating to the movement and momentum caused by the offset).

5) An elevated, ideal internal state of calm and composure, capable of an ethical influence of the flow of the interaction, while properly governing one's use and degree of force.

∽

Commentary on the Subtle Dimensions of Offsetting

It is easy to conceptualize the Type 1, 2 and 3 offsets with the themes we have chosen to describe them, Oncoming Traffic, Earthquake and Missing a Step. They were chosen specifically to assist the learning and teaching process by presenting a strong mental visual. The subtle dimensions of offsetting however, while being able to occupy the same categories as Type 1, 2 and 3 offsets can be quite different in their more subtle characteristics.

The essence of higher level offsetting, requires contentment with much more subtle use of energy. The Type 2 offset for example, instead of being a violent, explosive earthquake, can be rather a small adjustment, disrupting the recipient's balance just slightly, while still forcing them to adjust, and then borrowing that small amount of movement to gain momentum.

While in the ebb and flow, with momentum, a Type 1 offset, instead of the violent image of being struck by a car going 100 km per hour, we can imagine a more subtle constant of forward pressure, being assisted by momentum, such as when we successfully push a car out from being stuck. Both of these visuals are utilizing a Type 1 offset, while being distinctly different in the expression of it.

On the other hand, in the flow of energy, we must realize that the Type 1, 2 and 3 offsets are often not stagnant; they morph into each other in a manner almost imperceptive to the uneducated eye. It is at this degree of functionality that the more profound uses of offsetting come to life, as we draw less and less energy from our bodies' brute forces and in its place, generate movement from the subtleties of leverage, momentum and relationship to position and lines.

To do this well, requires also, a commitment to a controlled, inner condition, relying on attributes such as calm, composure and perceptiveness.

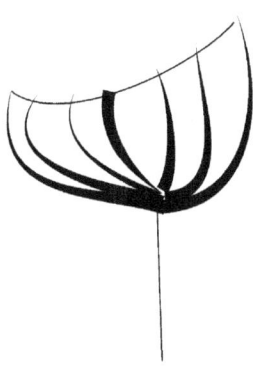

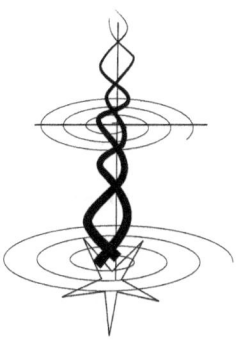

Chapter 3: And Furthermore

What makes close quarter martial technique work?

Everything that an opponent is able to do depends upon their own degree of balance. With good balance, good footing, whether mobile or stationary, that individual is much more able to disrupt you, if they have control over their own center of gravity.

If they do not have control over their own balance, they will innately be occupied with the need to regain it. It is this moment in time when they are occupied with regaining their balance that allows us to employ techniques according to our own intention more effortlessly. In addition, it is these moments, when our opponent's/partner's balance is offset, that we may also do nothing, as the offset is, all on its own, sometimes a sufficient degree of control.

This is why a well initiated attempt (from our opponent) at anything, such as a strike or push, is nullified (at least partially) if their sense of balance is upset. To me, in the physical sphere, offsetting is the foremost key.

In addition, because offsetting as a high skill cannot be confined to mere outbursts of raw strength, it becomes increasingly more important to develop the ability to read pressure, energy, intention and gain sensitivity to the lines of movement. When you combine these skills with an ideal internal state, consisting of a minimum of calm, composure and proper perception, then we are much more able to act appropriately.

When I say act appropriately, I mean to employ the right tactics for the moment, or draw from the right arts. Striking is not always the right answer to a situation, nor is locking, so we must be able to adapt, harmoniously according to justice.

∼

Common Pitfalls?

What are common pitfalls in martial arts?

This is a very important question. I am sure there are many different answers to this question. If I was going to narrow it down to one area, I would say we often spend an enormous amount of physical energy and time training our bodies and technique, but fail to consider the higher qualities of our nature. I personally feel drawing from our inherent spiritual qualities, even when expressing martial arts functionally, is a very important aim. So this leads us to ask, what does it mean to express spiritual qualities?

Human beings hold within them the potential to utilize profound powers of observation, creativity, strategy, reflection, contemplation, understanding, knowledge, wisdom and investigation, and may potentially utilize these powers in a noble way. The virtues of nobility are rightly active when utilized in service to humanity; virtues such as love, peacefulness, forgiveness, mercy, justice, patience and selflessness.

In martial arts, we have a potential laboratory to practice and observe these good qualities in action or to discern if these good qualities are absent. One of the ways we may be able to realize if good qualities are absent, is if we identify what a manifestation of not so good qualities can look or feel like.

In use of force training, there are certain variables that can come up within us, such as:

Feelings of:

1) Anger

2) Temper Loss

3) Retaliation

4) Vengeance

5) Desire to Display

6) Misplaced Competitiveness

7) Prejudice

Even misplaced or ill-timed expressions of the stress response (fight, flight or freeze) can reduce our abilities to act according to a higher ethic or more refined norms of our character and behavior. These are just some of the examples we can mention. These feelings are in a way an absence of a true quality or virtue, such as patience, forgiveness, compassion, calm, composure and awareness. When we feel the lower, the higher virtue is by default, absent.

In training, it is important to be able to gain an objective ability to identify lower qualities when or if they feel like they want to come up within us. In addition, we can reflect upon what are the best ways to progress internally, to draw from higher attributes, even when under the stress of training and daily life.

Part of the process of identifying higher or lower qualities within us, is to meditate, ponder and reflect upon them. Meditation has been a part of martial and yogic sciences for many years now. To develop the inner in service to the outer has been stimulated progressively throughout the ages by the stirring influences of our world's great Spiritual Luminaries, inspiring the penetrative and productive thoughts that reveal the profundity of science, art and philosophy. Adding meditation, contemplation, reflection and deeper thinking to our practice cannot but help us in our training and in our daily walk of life.

Meditation, when coupled with control and mindfulness over one's breathing, can be instrumental in gaining the power to better control our emotions, which in turn affects positively our responses to triggers that can cause expressions of misplaced anger or temper loss.

Managing these impulses is essential to growth as a human being. As we are social beings, we are designed to progress not in isolation from one another, but in unity and concord.

When we strive regularly and diligently to obtain an inner state that is calm, composed, mindful of our thoughts and actions, we can begin to adapt this inner condition, in a manner that animates positively the activities and interactions of our life.

Human beings hold the innate ability to ponder, reflect, investigate and understand. Martial arts can therefore be further galvanized to serve the cultivation of a peaceful warrior, manifesting fearlessness in the path of truth, forging the spirit of peacemakers and protectors of true civilization.

The future of the martial path is essentially set out before us according to our own potential advancement, which in turn means, we are ultimately responsible for the direction of our life's activities, taking account of it and making choices that are able to serve a higher good.

Though we have gained much from the martial sciences passed down to us throughout the ages, the future, potential refinement and direction of martial arts is in my opinion still yet to be discovered and unleashed. My feeling is, when we tap into the inner, spiritual susceptibilities latent within us, and learn to govern our actions in a refined manner that is cognizant of our personal, private lives and our shared collective lives, our martial arts will be transformed, evolving from a sword that kills, to a beautifully refined instrument of service and cultivation.

∼

Offsetting: Applied to Outside Ranges

When one intercepts (using hands or feet), they are in a sense, disrupting the recipient's center of gravity, by directly stopping **(a portion or the whole)** of the movement that generates the attack. When we stop hit, we are often temporarily jamming the recipient's ability to move and therefore disrupting their sense of balance. When cultivating the ability to intercept (while in outside ranges or within close contact), it involves gaining the perceptiveness to be able to disrupt an opponent's preparation, execution or intention. When we generate force, like unto a hit, or explosive push, we are simultaneously jolting and moving the recipient's center and this can be done destructively or nondestructively.

In essence, from the outside, it is possible to 'stop' the forward movement of those coming toward you, with hands or feet, thus jamming their center, which is the root of offsetting. Though we are discussing the empty hand, this principle resembles elements of both eastern swordsmanship and western fencing, when we gaze the sword mentally and physically towards, through and beyond the opponent's centre line and center of gravity. Though when we are in close quarters, or what I often refer to as the 'realm of cohesion', the dynamics of offsetting is more pronounced, the outside range is not confined to pain or injurious uses of force, but in fact can be utilized efficiently to cause, control, stimulate and relate to movement, all of which can function directly or indirectly in conjunction with the sciences to offset.

The outside ranges can also be highly developed in order to facilitate the entry into close quarters, to bridge the gap. This essentially means, the disruption caused by the interception, is a moment from which we may enter (or close the proximity) into close quarter range.

In a sense, we are learning to disrupt from the outside, and then enter into close range, in order to further disrupt or control. This can also be done in reverse, meaning we go from the inside range, while in close contact, to the outside, remaining attuned and responsive to the opponent's movements and intentions.

How Do We 'Provoke Movement, without 'Invoking' Escalation?

This book is to a certain extent dedicated to this question. Yet, I would like to take a moment to answer it more directly, as it is a question that I have been asked frequently over the years.

In short, it became my first priority in development to be able to provoke movement, without triggering an escalated reaction in the recipient. I spent and continue to spend enormous hours connected to others physically in practice. By spending the time, we gain the necessary laboratory from which to examine the results of our attempts.

If we already possess a habit of muscling our way through things, then we will need to pull back on that. This may take a complete transformation from within (as it did me), for when we give up muscling everything, we will experience a sense of loss. The reason for this is, we are no longer able to compensate (through brute force) for what we lack in our abilities, but will need to increase our reliance on momentum, leverage, timing, sensitivity, inner calm, composure etc.

When we have decided from our heart to give up the path of reliance on brute force and impulsive qualities, long term, for the higher road, we gain access to a different sphere of attributes within us. These are the qualities that were previously overthrown by misplaced competitiveness or the fear of losing. Of course, the process of gaining these attributes and qualities, can involve being thrown around a lot. But this is an important part of the process.

It becomes a path of discovering exactly what we do that triggers impulses and reactions in the recipient/opponent. For me, as I became aware of the impulsive reaction generated by muscled movement, I began to slowly, but surely, discover keys to moderating my use of force.

Three Keys to 'Moderating Force'

a. The first is the type that is used when the person you are attempting to move is bigger or stronger than you. In this situation, it is most common to act or react with brute force, which often proves to have serious limitations and repercussions.

The alternative is to use very small bumps and tips, choking it off. This is basically to cause small amounts of movement, without falling into excessive force. When we choke it off (limiting our own use and outbursts of energy), we do not need to sustain it. In turn, the recipient or opponent has very little to react to, as the force we are generating comes and goes too quickly.

With this (like moving a vehicle that is stuck), we learn to compliment the caused or existing movement, and add to it according to the presented momentum. In addition, 'riding the bull' exists here, as we cannot do this forcefully, you will need to go with the flow, so that you can generate/compliment the momentum.

b. The second type of moderating force is based in sensing that you are able to cause movement, without the need for excessive force. In this case, you must not give in to the potential inner suggestion to add force just because you can. To avoid the trigger means avoiding our own desire to use more force. We again must rely more on momentum, leveraging, while complimenting a caused or already existing direction.

c. Another very important key is to literally 'support' the recipient of our energy and force. The recipient should feel what we are doing to them is not based in aggression and that we care for their wellbeing. Think of this like we are playing with a child. When playing with a child, we may be in a sense manhandling them, yet doing so in a manner that does not induce fear or a lack of trust.

This must be progressively developed in our practice and is not purely technical. When we do this, it almost innately induces a feeling of trust in the recipient, as they should sense that we are moving them, yet doing so in a manner that is supportive of both their physical frame and their well-being.

I have often said that my 'touch downs' or my 'winning points' progressed to become much more subtle. I discovered that I gained more, by contenting myself with much less, even at times almost imperceptibly less. I now strive to pay attention to very subtle movement, rather than rely on big, large movements.

If you move someone, with small moves, rather than always a continuous 'forced' one, then the recipient has very little to respond to. What a recipient usually innately responds to is based in a sustained amount of pressure or force upon their body or limbs; therefore, we must not sustain our movements, but choke them off.

This book is sprinkled with new terminology regarding this, such as resistance chambering, where we let go of a chambered resistance, or shadowing, the ability to follow the flow of movement, without overly applying force along the way, etc.

The Need for Hands On Training and Further Study

I realize what I am attempting to communicate requires the addition of hands on training and visual aids. Learning from a qualified teacher is essential at some point, if one is serious about attaining these principles at the level of a high skill.

During my own process of training, I learned early on that 'feeling' the skills and qualities of those more experienced and of greater skill was extremely valuable. I would do everything I could to train 'personally', hands on, with the experts, regardless of whether they were known to the world or not.

It is important to gain the lasting impressions that can only come from feeling high skill and attributes upon your body. These experiences are capable of causing us to contemplate and strive to emulate these feelings over the years to come.

In addition, one of the reasons I presented a list of martial arts distinctions was to hopefully help some individuals realize that not all expert martial artists must look good to the eye, as they can be perfectly great at what they do, whether an onlooker is aware of their skill, impressed by it or not.

For the most part what I am describing in this book, even when done and witnessed at the higher levels is not necessarily going to fulfill anyone's need for martial arts eye candy. Martial arts, largely due to the appeal of action movies and entertainment, have often been boxed into needing to fulfill the entertainment and esthetic based desires. Offsetting can be a truly subtle science, and it is best appreciated when understood and personally experienced.

Chapter 4: Close Quarters Arts and Distinctions

When interacting physically within close quarters there are several areas that we like to isolate and flow within while upright. Each of these areas provides a framework from which offsetting can occur.

1) **Conventional Clinch:** The conventional clinch is when we clasp both hands around the recipient's neck, while simultaneously pulling and twisting from the wrists, forearms and palms. The elbows often occupy centerline directly, being placed down the center of the chest and sometimes just off of it.

    The energy being utilized is based in strong forceful bursts of motion to 'offset' in a dominantly 'Type 1' manner. The physical characteristics of a fully engaged conventional clinch are usually tight.

2) **Modified Clinch:** The modified clinch rather than being as tight as the conventional clinch is now characterized by being open, with more space, thus allowing room to utilize different techniques, qualities and attributes.

    While holding the neck, our limbs begin to spread out, occupying the chest, collar bones and the shoulders. The limbs also become more mobile, allowing for more options to manipulate the recipient's neck, chin, shoulders, chest, and limbs, while remaining in contact.

The modified clinch contains the potential to be extremely explosive and percussive as it has more room and is less tight in its structure. The modified clinch flows often between Type 1 and 2 offsets and can also take on very subtle characteristics, more so than the conventional clinch.

3) **Circle of Space:** The circle of space is characterized by its detachment from the back of the neck, yet remains in continued relationship to the recipient's body, as well as the base of the limbs. The limbs are beginning to be utilized to cause offsets and more subtle qualities are being employed regularly. Type 1, 2 and 3 offsets are possible with a harmonious interplay. The space in between both parties, I often compare to an open umbrella that has been spread out between the centers of both individuals.

4) **Flow Hands 2:** Flow hands become active when the limbs are still in contact, yet not engaged in any of the clinches, but in cohesion with the limbs. Flow hands utilize the pressure reading, sensitivity and line familiarization gained from sticking hands, pushing hands, passing, pulling hands, and percussion, which all work together potentially in a continuum of transitions and flow.

Though this area of training needs to be taught and experienced hands on, I will offer some thoughts here. I often see flow hands being utilized without sufficient pressure and energy, which causes a complete disconnect to the partner's center of gravity.

Flow hands should be dynamic, but not always co-operative, at least not as we mature into it. These training methods can fool us into being overly co-operative, due to it starting out with the need for both parties to 'learn' basic flow patterns.

The structures and patterns of play in flow hands training, should eventually be transcended, yet remain in a manner that preserves the structures in order to 'return' to them, rather than rely upon them as the sole extent of its possible expression.

If the path is traversed deeply enough, at least one person can dominate this area through potent offsetting (by constantly connecting to the recipient's/partner's center of gravity), combined with dynamic transitions and high level reading of pressure. When one is able to command this area and flow in an out of the areas listed above, the power of offsetting will reveal itself in splendour.

When we spend the time, the ability to offset increases through subtle manipulations of the limbs. This is done in an almost imperceptible adjustment which lines up nicely with the recipient's center of gravity, hence activating the offset.

Flow hands, while starting out with a pattern of drills, can eventually become a flow of lines and contact that does not depend upon 100% co-operative drilling. If flow hands training is understood and practiced properly, it will produce a much higher level of functionality while in close quarters.

5) **Shifting Arts:** The first science of offsetting occurs in the shifting arts, the arts and sciences of moving a person's body weight. This art can become a very sophisticated element of martial prowess, for it is simple and effective, while being highly absorbable during the initial learning process. It can be learned easily and mastered over one's lifetime with profound results.

The shifting arts involve all of the components mentioned within this book regarding offsetting, including the above conventional clinch, modified clinch, circle of space and flow hands; it can also can take on technical variations when isolating a limb or upon the recipient's body, when manifested through the following classical arts: the locking, trapping, and striking arts.

6) **Locking Arts:** The second science from which we may employ 'offsetting' is in the locking arts. Locking is essentially pressure applied against the joints and digits. The locking arts are a very high level science, born out of the merging of intelligence and ethics, as it was developed in order to be able control harmful individuals without the need of destructive or lethal force.

In our system, we discern between four distinct expressions of the locking arts, being:

1. **Entry Locks:** to cause movement
2. **Restraint Locks:** to control movement
3. **Compliance Locks:** to control movement through pain compliance
4. **Conclusive Locks:** to cause injury to the joint, ligaments, muscle etc.

To utilize the locking arts for the purpose of offsetting requires we achieve expertise in the 'entry' locks. The purpose of the entry lock is to cause movement, a type of movement that is assisted by momentum, rather than pure pain compliance.

To employ an entry lock, is to learn to distribute energy and pressure into the joint, while simultaneously dispersing it throughout the limb, connecting it to the recipient's center.

7) **Trapping Arts:** The third science from which we may employ offsetting lies within the trapping arts. Trapping is basically the ability to 'immobilize the recipient's limbs'. In doing so, one may then be able to employ shifting, locking or striking, in conjunction with trapping.

High level trapping is often characterized by its percussive nature and by its need for advanced sensitivity to pressure and energy. It is an art that requires mastery and once achieved, is an important component to the art of offsetting.

Trapping is also a unique practice, as one can obtain proficiency in what I have termed a 'pressure transfer'. A pressure transfer essentially means to move from one point of pressure, to another, seamlessly, and without any dislocation, such as from one limb to another.

The subtle dimensions of offsetting can be implemented through the assistance of trapping, for when we have immobilized a limb, we may also disrupt the recipient's body weight, forcing them to adjust. This is done very indiscriminately at the higher levels of skill. Trapping may be utilized in conjunction with a strike or explosive push, disrupting the recipient's center of gravity abruptly, while immobilizing the limbs.

Trapping is an art form that requires deep study, practice and contemplation, for it is a fine art, requiring refined qualities.

8) **Striking Arts:** The fourth science we may employ in conjunction with offsetting is striking. Striking can be utilized from within close quarters as well as from outside of close quarters. The most options for striking exist within close quarters.

When considering 'offsetting' with respect to the striking arts there are two basic premises from which we may choose to strike:

The First: We strike during the utilization of offsetting. This means we employ a strike 'while' the person's balance or equilibrium has been upset. Striking can be particularly potent when employed this way, as it is essentially two goals working in harmony, one is to achieve the offset and the other to strike.

The Second: We strike during a moment when we are not in control. To me, being in control means I have the ability to offset, to control a person's center of gravity. When I am unable to do this, I sense it; this in turn may activate the need to strike, in order to regain this control. Acting on this or not is based on factors of appropriateness, measured within the flow of circumstances.

∽

A Note:

During the initial years of training in the conventional clinch, I was fortunate to have partners that were skilled and sometimes stronger than I was. After a time, I began to realize how much muscling into a superior position occurred within what I now term, the conventional clinch.

I decided to conduct my own research and development into how to achieve a more effortless ability, while in the conventional clinch. I learned that I literally had to 'turn off' my desire or 'habit' of matching the common tendencies of that range.

I began to sacrifice the common impulses that result from activating brute strength and muscled manipulation. In turn I invited characteristics found within what I would now call 'Riding the Bull'.

This process of learning to read the energy being applied to me, and in turn, implementing characteristics such as complimenting, movement to momentum, choking it off and shadowing, led to the subtle range distinctions within close quarters, namely modified clinch and circle of space.

The discipline of doing this, literally called for me to create a new mind, one that was reliant on subtle qualities, such as being calm and not concerned with feelings of competitiveness that can occur in the conventional clinch.

If I became forgetful of these subtle characteristics, I would find myself easily falling back into the common mode of doing the conventional clinch. This process of learning was very valuable to me. I was able to acquire a completely new set of skills and qualities than I previously had.

I have come to realize that when you have attached yourself (your own center) to the recipient's center of gravity, moving them can become like moving your own body weight.

Chapter 5: Applied Jeet Kune Do Philosophy

Jeet Kune Do is my most prominent martial foundation. I have been a dedicated student and teacher of this art, for more than 20 years. An aspect of its philosophy I have often revisited is the concept of stages in one's process of acquiring mastery in the martial sciences.

As offsetting is one of the focal elements in the ability to express high martial skill, I felt discussing elements of the philosophical premise of stages in one's cultivation would be a relevant offering.

~

Stages of Cultivation

There are three Stages of Cultivation in Jeet Kune Do. These stages mirror the process of learning and skill development originally fleshed out through the contemplation of Chan (Zen) Buddhist and Taoist precepts respectively.

Within the philosophy of Jeet Kune Do, which borrows and applies heavily the universal precepts of Buddhist, Confucian and Taoist teachings, as well as scientific analysis, there are certain guidelines and principles that organize and assist in developing the accuracy of our intuition. This enables us to direct our acquired skill sets to 'fit in' appropriately, within a changing moment.

These guidelines and principles are meant to be presented in an intelligent, progressive breakdown, which exposes the keys to unburden the mind, pointing to the gateway of the third stage in Jeet Kune Do training, the tranquil 'stage of emptiness'.

∼

Stage 1: Partiality

In Jeet Kune Do, there are three stages we must strive to traverse. The first is called 'the stage of partiality'. The partiality stage consists in the systematic breakdown of our primitively based naturalness and impulsive bodily behaviours. Simultaneously, we begin moulding and fashioning ourselves in order to formulate the building blocks of higher attributes, techniques and perceptiveness.

This phase is for many, one that is never fully traversed. This may be due to a lack of proper guidance or to one's 'inability' to 'empty their cup'. In the long run, a mature practitioner may often return here for the reason of maintaining acquired skill sets, as well to more deeply assimilate the necessary evolution of their skills or to add completely new ones.

Individuals of the art that return to partiality do not experience this phase in the same manner as they did in the beginning. A mature practitioner, or one having already experienced the third stage of emptiness, revisits the first stage with a new, heightened state of flow and learning absorption. These individuals are already comfortably familiar with the necessity of overriding one's lower impulses and the need to empty one's cup.

∼

Stage 2: Fluidity

The second phase is the 'stage of fluidity'. When they reach this phase, many also remain here, fluctuating between stages 1 and 2. Stage 2 can offer a feeling of functionality and the pride of applicable knowledge yet is still confined by the 'analytical mind'.

The analytical mind is a naturally human condition, occurring when we become burdened by a constant overanalysis of our supposed knowledge and its application.

Yet precious time, energy and ability are sacrificed when clinging to this supposed knowledge. If this mistake is not eventually detected, this stage may confine the practitioner, for a very long time. In truth, many seasoned practitioners, unaware, never truly break out of this phase.

This stage, of being burdened by the mind, affects negatively one's ability to act according to the efficient flow of an acquired skill. Those that are within this stage may in fact know a lot of techniques, understand martial art intellectually, and yet regardless, lack the required transcendence of this knowledge to truly express their skills, unobstructed.

∽

Stage 3: Emptiness

The third stage is the 'stage of emptiness'. This is the highest stage in Jeet Kune Do, as it is the place within ourselves where we are able to act without any hindrance from the mind or the body's lower impulses.

It is the most tranquil and functional stage of the three, and is the portal towards even greater accomplishments. Within this stage, the body relies more upon the spiritual attributes and qualities of a calm and serene mind, rather than reacting to the changes and chances affecting its temporal, physical condition.

When we have reached the third stage, we have thus lessened the pull of our natural, lower impulses and reside in a tranquil, yet 'pending state' of awareness, ability and even devastating explosiveness. Our skills have been honed accordingly, and have become beyond premeditation, seemingly endless analysis or the drone of physical repetition.

This understanding and acquired condition makes it possible to be truly simple and direct, as our mind becomes accustomed to moving according to the circumstances. When the body has been sufficiently trained and responsibly maintained, it acts efficiently according to the need.

∼

The Fitting in Spirit

Though every circumstance is unique, the 'essence' of these circumstances is both identifiable, possible to be organized, and therefore prepared for. It is this key point that makes this wonderful philosophy a scientifically bound one, yet universal in its potential application.

The portals provided during learning and training can eventually lead to the 'stage of emptiness'. This lofty, yet attainable goal however begins first with understanding the basic insight of distinctions in range, conditions and proximity.

In a comprehensive Jeet Kune Do study, which is beyond the scope of this particular book, the ranges of hand to hand interactions are broken down, organized and meticulously refined, first by their common natural distinctions, then by the ease of intelligent strategies that may be discovered and expressed according to one's capacities.

This process is then furthered according to one's personal needs, standards, abilities, strengths or weaknesses. It is this element that makes Jeet Kune Do a 'potentially' personal art, yet I feel the practitioner must traverse it in its entirety to taste of the ripest fruits of one's labours.

∼

Stages of Cultivation Applied to Offsetting

Offsetting is primarily a close quarter skill. Within this book, we have provided some distinctions of range, characteristics and proximity, applied to offsetting. It is an element that is considered foremost in our tactics, as it sets the tone for our choices within the moment that a successful offset provides.

By mere observation of an 'Interaction or Altercation' it is clear that it is not confined to close quarters. This is why the science of offsetting as broken down within these pages, is a mere portion, though I feel an extremely important one, of what is potentially contained within the whole process of training or the play out of an interaction.

Offsetting as presented within this book is possible to be reached at the third stage of cultivation, which is to bring our abilities and awareness to an unburdened sense of freedom, while acting appropriately. I feel it is not only possible to reach the third stage, but within our capacity as a species to do so.

Offsetting is not bound to represent any style, martial art, martial system or founder. With this said, it is my hope that we examine offsetting according to the full scope of its possibilities, in light of all the elements of our inherent noble, inner and outer potentialities.

Closing Comments

This book acts as the first in a series of three books. It was purposed to be not a large work, but part of a series, capable of acting as both a study tool and an analysis of the roots of the martial sciences as I understand and strive to present them.

It was intended to be presented not as a breakdown of a particular style or martial arts form, rather as an analysis of an essential root in martial skill, which by its own nature is not bound to represent any current or past martial art, system or founder. It is however a science, poised to advance forward, perhaps able to provide a stepping stone to further develop the use of force arts, as we are collectively destined to grow in capacity, as we mature into an enlightened humanity supportive of a peaceful world embracing culture.

The entirety of the book series will introduce applied sciences of offsetting, considering factors such as personal protection, professional use of force, alongside an in-depth analysis of offsetting when considering classical elements of martial skill. The book series will also delve deeper into the consequences of one's inner condition upon the science of offsetting and our performance in martial art.

It is my hope that this book, 'An Introduction to the Science of Offsetting', will be utilized by the reader, in conjunction with the two books following in the set as they are made available.

It has been a great pleasure for me to write this book. I have high hopes for its readership and its use.

I hope you enjoyed it and please do not hesitate to contact me personally.

In Peace

Jamen Zacharias

The Symbol of Circles and Lines

Our symbol for offsetting in the front cover image represents two prominent aspects within the external and internal spheres, being: Linear Energy & Circular Energy.

The two crossing lines deep within, represent a connected, linear movement as well as the center of gravity.

The circles represent a flow of movement that we impose upon the recipient, moving them out of their center, in various directions and in various modes, while maintaining our own balance, internally and externally.

The circle that encompasses all of it is the 'center' from which we must act. This center rests intuitively and in balance ideally from an inner place of deep calm and peace.

This symbol also aims to celebrate the harmony of two important facets of human life; the actualization of an ever maturing scientific process, with that of our spiritual capacities and noble aim.

The two lines going into the circle represent a lifelong path, a process from which we strive to manifest the art and sciences in a manner noble and good.

About Jamen Zacharias

I began my journey in martial arts when I was very young. In the early '90s I found the art that would become my foundation, Jeet Kune Do (founded by Bruce Lee), eventually becoming certified as a Senior Full Instructor.

Alongside this, I studied with vigor the arts of edged and impact weapons, based out of the Philippine martial art of Kali (Arnis, Eskrima). In addition, I ventured into Japanese and Korean Swordsmanship, alongside my continued practice of the Philippine martial arts and Jeet Kune Do.

In my initial years, I worked hard to acquire the tools I felt necessary to obtain efficient, potent street fighting abilities. This process led me to learn from some of whom I and many others consider, the very best Hand to Hand Combative Trainers.

Some of these teachers are contracted regularly to pass on knowledge to the highest levels of specialized military groups and law enforcement. I still consider it not only an honour to have learned directly from these men and women, but absolutely essential to my martial arts process.

Since my early teens, I have struggled against a chronic illness called Crohn's disease (Inflammatory Bowel Disease) and after a series of life saving surgeries, I ended up with a permanent ileo-ostomy. During the most trying years of my early life, I decided from deep within to pursue the lifelong path of martial arts mastery and strive to become a qualified teacher.

In 1999, after much search and investigation into the purpose and reality of religion and spirituality, I entered the Bahá'í Faith. From that moment, a shift occurred inside me to search deeper into the arts and sciences, to discover the root and to glimpse into its essence, so that I may begin to understand how force may be utilized in service to justice and nobility.

With this said, I offer this book as a mere beginning, not as an end, of my own learning. It is with great pleasure that I offer this book to you, the reader.

Certifications:

Jeet Kune Do
1. Senior Full Instructor of Jeet Kune Do – Thomas Cruse
2. Full Instructor of Jeet Kune Do – Michael Gruener

Edged & Impact Weapons-Craft
1. Senior Full Instructor of Filipino Martial Arts – Thomas Cruse
2. Full Instructor of Filipino Martial Arts – Shishir Inocalla

3. Full Instructor of Filipino Martial Arts – Michael Gruener
4. Edged Weapons Instructor – Paul Vunak
5. Black Belt Instructor of Arnis/Tai Chi Gung – Shishir Inocalla
6. Advanced Certificate Edged & Impact Weapons – Jerson Tortal

Swordsmanship
1. 2nd Degree Black Belt in Iaido – Mr. Jong
2. 1st Degree Black Belt in Kumdo – Mr. Min

Notes

Notes

Notes

www.ingramcontent.com/pod-product-compliance
Lightning Source LLC
Chambersburg PA
CBHW071334190426
43193CB00041B/1772